How to be Brilliant a

NUMBERS

Beryl Webber
Terry Barnes

We hope you and your class enjoy using this book. Other books in the series include:

Maths titles

How to be Brilliant at Algebra	1 897675 05 4
How to be Brilliant at Using a Calculator	1 897675 04 6
How to be Brilliant at Shape and Space	1 897675 07 0

English titles

How to be Brilliant at Writing Stories	1 897675 00 3
How to be Brilliant at Writing Poetry	1 897675 01 1
How to be Brilliant at Grammar	1 897675 02 X
How to be Brilliant at Making Books	1 897675 03 8
How to be Brilliant at Reading	1 897675 09 7
How to be Brilliant at Spelling	1 897675 08 9

Science title

How to be Brilliant at Recording in Science	1 897675 10 0

If you would like further information on these or other titles published by Brilliant Publications, write to the address given below.

Published by Brilliant Publications,
PO Box 143, Leamington Spa CV31 1EB

The publishers and authors are grateful to the pupils and staff at Cherry Orchard School in Birmingham for their help in trying out the activities.

Written by Beryl Webber and Terry Barnes
Illustrated by Kate Ford and Tony Dover
Cover photograph by Martyn Chillmaid

Printed in Great Britain by the Warwick Printing Company Ltd

ISBN 1 897675 06 2

First published in 1995
10 9 8 7 6 5 4 3 2 1

Contents

Introduction

How to be Brilliant at Numbers contains 42 photocopiable sheets for use with 7–11 year olds. The ideas are structured in line with the National Curriculum programmes of study. They can be used whenever the need arises for particular activities to support and supplement whatever core mathematics programme you use. The activities provide learning experiences which can be tailored to meet individual children's needs.

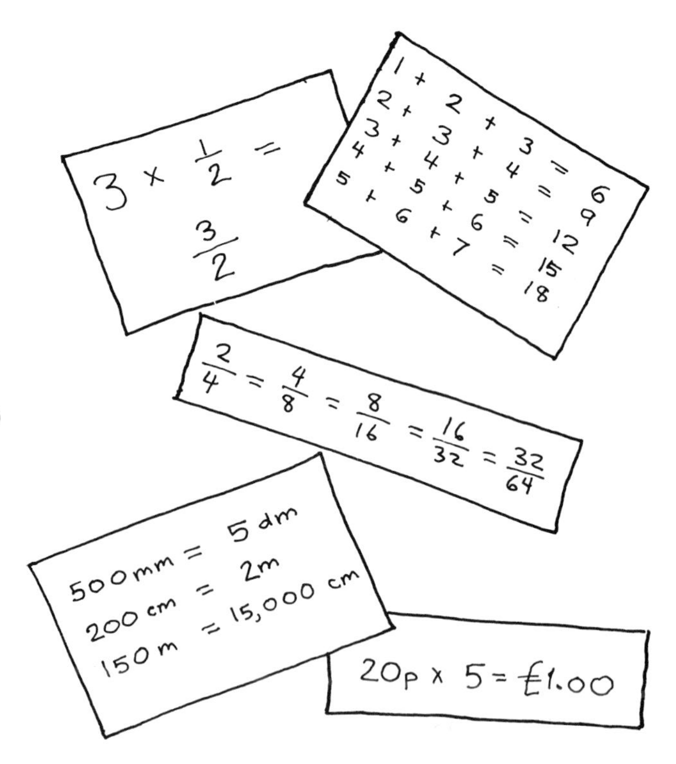

The activities are addressed directly to the children. They are self-contained and many children will be able to work with very little additional support from you. You may have some children, however, who have the necessary mathematical concepts and skills, but require your help in reading the sheets.

The children will need pencils and should be encouraged to use the sheets for all of their working. Some activities require extra resources such as coloured pencils, scissors and a calculator. Some will require the use of additional resource sheets and these can be found at the back of the book. Where this is the case, it has been indicated by a small box, with the relevant page number in it, in the top right corner, eg 47.

How to be Brilliant at Numbers relates directly to the programmes of study for using and applying mathematics, number and data handling. The page opposite gives further details and on the contents page the activities are coded according to programme of study and difficulty. The level of difficulty is indicated by a letter code (A-C) and is provided to give you an indication of how the activities relate to mathematical progression within the key stage. Activities coded 'A' are the most challenging.

Page 44 provides a self-assessment sheet so that children can keep a record of their own progress.

Links to the revised National Curriculum

The activities in this book allow children to have opportunities to:

- use and apply mathematics in practical tasks, in real-life problems and within mathematics itself;
- take increasing responsibility for organizing and extending tasks;
- devise and refine their own ways of recording;
- ask questions and follow alternative suggestions to support the development of reasoning;
- develop flexible and effective methods of computation and recording, and use them with understanding;
- use calculators to enable work with realistic data;
- consider questions using statistical methods.

In particular these activities relate to the following sections of the Key Stage 2 programme of study.

Using and Applying Mathematics

2. Making and monitoring decisions to solve problems

a select and use the appropriate mathematics and use the appropriate materials;

b try different mathematical approaches; identify and obtain information needed to carry out their work;

c develop their own mathematical strategies and look for ways to overcome difficulties;

d check their results and consider whether they are reasonable.

3. Developing mathematical language and forms of communication

a understand the language of:
- number
- relationships, including 'multiple of', 'factor of' and 'symmetrical to';

b use diagrams, graphs and simple algebraic symbols;

c present information and results clearly.

4. Developing mathematical reasoning

a understand and investigate general statements;

b search for pattern in their results;

c make general statements of their own, based on evidence they have produced;

d explain their reasoning.

Number

2. Developing an understanding of place value and extending the number system

a read, write and order whole numbers, understanding that the position of the digit signifies its value, use their understanding of place value to develop methods of computation, to approximate numbers to the nearest 10 or 100, and to multipy and divide by powers of 10;

b extend their understanding of the number system to negative numbers, in context, and decimals, in the context of measurement and money;

c understand and use, in context, fractions and percentages to estimate, describe and compare proportions of a whole.

3. Understanding relationships between numbers and developing methods of computation

- **a** explore number sequences, explaining pattern and using simple relationships; interpreting, generalizing and using simple mappings;
- **b** recognize the number relationship between co-ordinates in the first quadrant of related points on a line;
- **c** know addition, subtraction and multiplication facts; develop a range of mental methods; use some properties of numbers;
- **d** develop a variety of mental methods of computation;
- **e** understand multiplication and division; use associated language and recognize situations to which these operations apply;
- **f** understand and use the relationships between the four operations, including inverses;
- **g** extend methods of computation to include addition and subtraction of negative numbers, addition and multiplication of decimals, calculating fractions and percentages of quantities.

4. Solving numerical problems

- **a** use the four operations to solve problems;
- **c** check results by different methods.

Handling Data

2. Collecting, representing and interpreting data

- a. interpret tables used in everyday life; interpret frequency tables;
- c. understand and use measures of average, leading towards the mode, the median and the mean in relevant contexts, and the range as a measure of spread;
- d. draw conclusions from statistics and graphs, and recognize why some conclusions can be uncertain or misleading.

Each activity has been coded on the contents page to indicate its main relationship with the above aspects of the programme of study for Key Stage 2.

The coding operates as follows:

N – Number
HD – Handling Data

These letter codes are followed by a number and lower case letter to indicate the relevant sub-section and aspect.

For example:
HD2(c) indicates Handling Data, sub-section 2 (Collecting, representing and interpreting data), c – 'understand and use measures of average…'.

Each activity also relates to the Using and Applying Mathematics section of the programme of study in a variety of ways.

Each activity is also coded by an upper case letter (A-C) indicating the relative difficulty of the activity itself. Activities coded 'A' are the most challenging.

Ancient Greek numbers

The ancient Greeks used a different number system from ours.

They used these symbols for the numbers 1 to 10.

A	B	Γ	Δ	E	F	Z	H	Θ	I
1	2	3	4	5	6	7	8	9	10

They had no symbol for zero and they had different symbols for the tens numbers.
For example:

N is the symbol for 50, and

Π is the symbol for 80.

This made adding and subtracting very hard.

Write the Greek numbers for 11–19, 50–59 and 80-89.

Do you think the order of the symbols mattered?

It does matter for our symbols. 31 is not the same as 13.

EXTRA!
Investigate some facts about Greek, Roman or Egyptian numbers.
Find out where the symbol for zero came from.

Make twenty

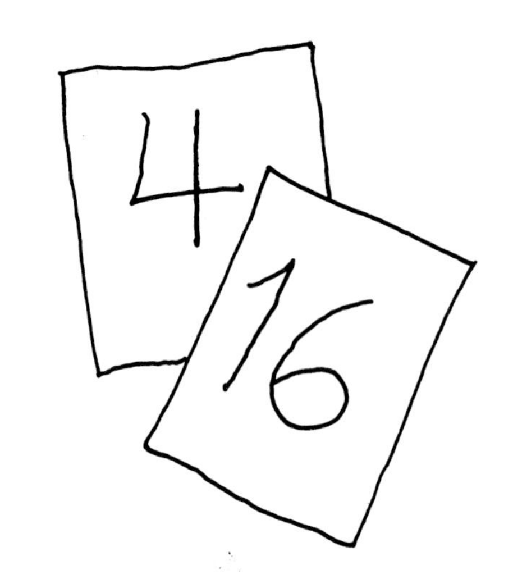

You will need two copies of the Numbers resource sheet (page 45) and two friends.

Cut out the number cards and shuffle them.

This is how to play:

- Deal all the cards to the three players.
- Each player secretly looks at his own cards and lays down any pairs that total 20 face up on the table.
- Player 1 holds up his remaining cards – keeping the numbers hidden.
- Player 2 takes one card. She looks to see if it makes a pair that totals 20 with any card in her own hand.

  If it does, the two cards are placed face up on the table and Player 2 has another go (with Player 3).
- If it does not make a pair, Player 2 adds the card to her hand and then Player 3 chooses a card from Player 2's hand.
- The winner is the first player to have no cards left.

EXTRA!
Find ways of making 20 with three cards.
What about four or five cards?

Coins, 1

You will need lots of 2p, 5p and 10p coins and seven different coloured pencils or felt-tip pens.

1	2	3	4	5	6	7	8	9	10
11	12	13	14	15	16	17	18	19	20
21	22	23	24	25	26	27	28	29	30
31	32	33	34	35	36	37	38	39	40
41	42	43	44	45	46	47	48	49	50
51	52	53	54	55	56	57	58	59	60
61	62	63	64	65	66	67	68	69	70
71	72	73	74	75	76	77	78	79	80
81	82	83	84	85	86	87	88	89	90
91	92	93	94	95	96	97	98	99	100

Tip: Be careful when circling the numbers. Some will end up with five different circles!

- Choose one coloured pencil and circle all the amounts you can make using only 10p coins.
- Choose a different colour and do the same for 5p coins.
- Repeat for 2p coins with a different colour.
- Now investigate the amounts you can make using at least one 10p and one 5p coin, but no others. Circle these amounts using a fourth colour.
- Try with 2p and 10p coins. Circle the amounts using the fifth colour.
- Repeat for 2p and 5p coins. Circle the amounts with the sixth colour.
- Which amounts can you make using at least one 2p, 5p and 10p coin? Circle the amounts in your final colour.

EXTRA!
Which amounts can't you make?
Which amounts can be made in the greatest number of ways?

Coins, 2

You will need lots of different coins.

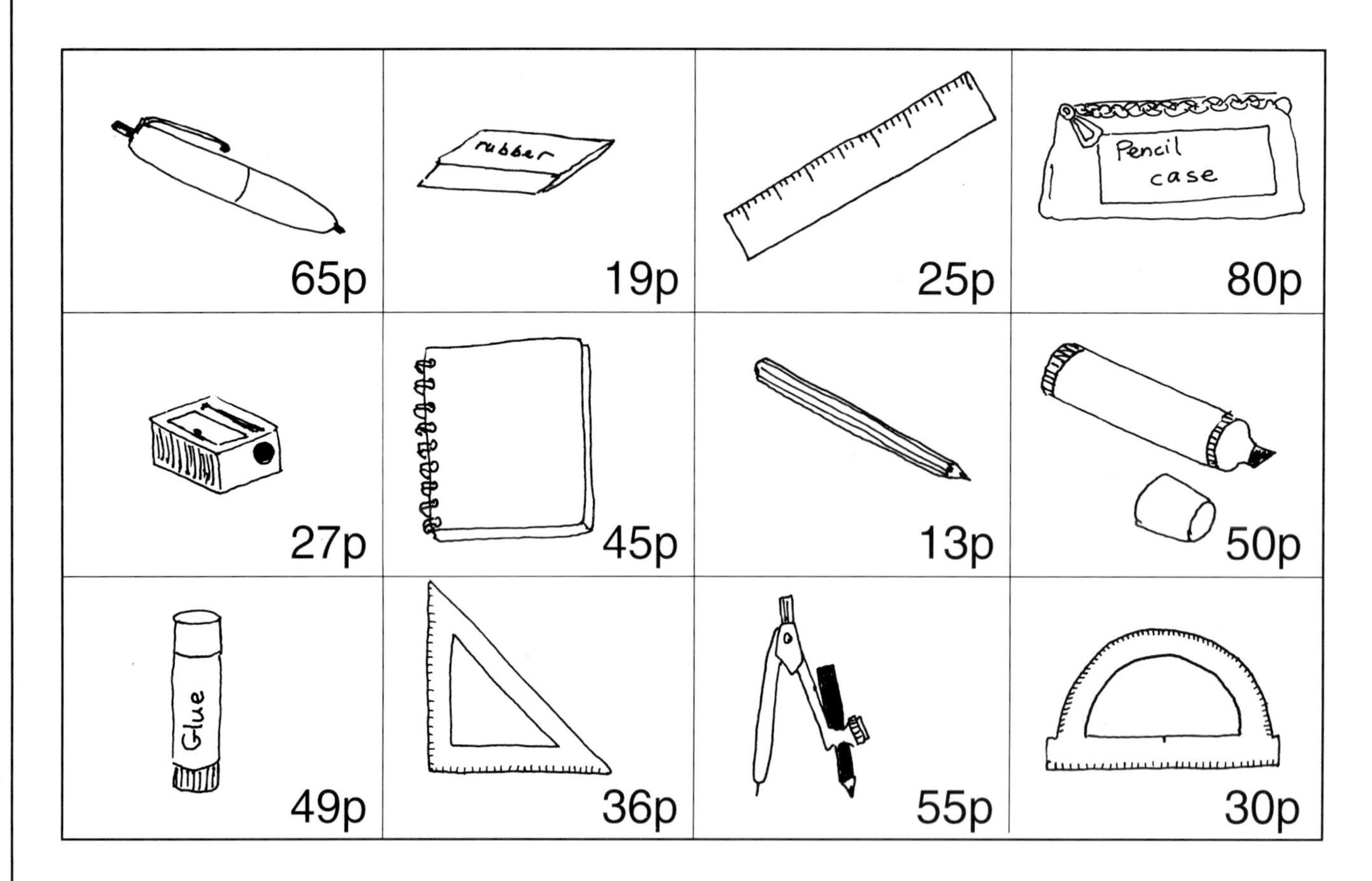

- Which item(s) can you buy using only one coin (without having any change)?
- Which can be bought using any two coins?
- Which can be bought using any three coins?
- Which item needs the biggest number of coins?

EXTRA!
Which items could you buy with a number of coins and receive only one coin as change?

For example:

Notebook: 45p → pay   

and receive [5p coin] as change.

Spend £5

You want to buy at least one of each type of plant.

You have £5 to spend. How will you spend it?

Plants may be bought singly.

Tray of pansies
£3.00 per tray of 12

Strip of marigolds
£2.40 per strip of 6

Pots of sweetpeas
£1.00 for 4

Pots of miniature roses
£13.50 for 9 pots

EXTRA!
Investigate how many of one type of plant you could buy for £5.

The garden centre

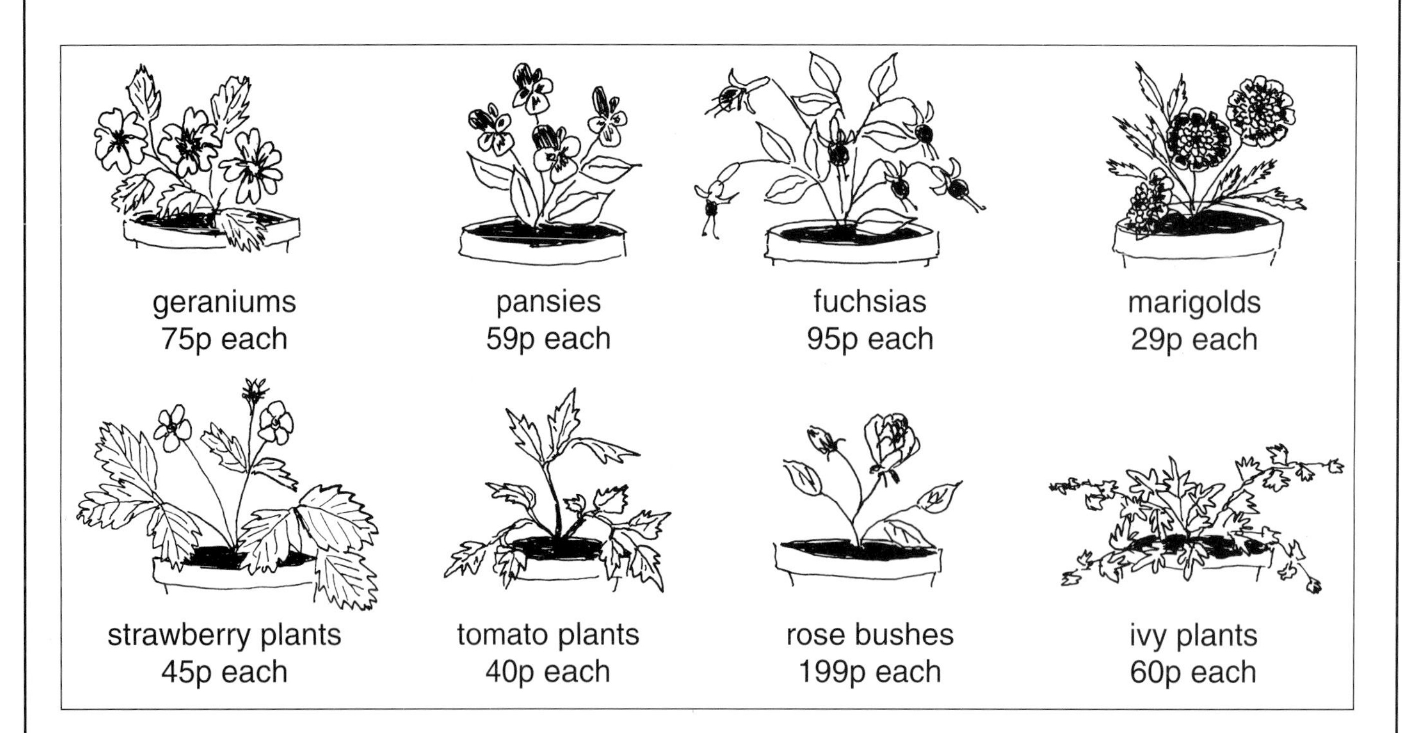

A school collects £150 to buy plants for their garden. They have already decided that they want:

- at least 15 rose bushes;
- at least 24 tomato plants;
- 10 large pots of flowering plants with 15 plants in each;
- strawberries to sell at their summer fete, so they need at least 10 plants;
- a flower border if there is any money left.

Decide which plants they should buy, remembering the decisions the school has already made.

Tip: Calculate the cost of the roses and tomatoes to find out how much is left for the flowering plants and strawberries. Then decide which plants to have in the pots, but make sure you have enough money left for at least ten strawberry plants. Any extra money can be spent on plants for the borders.

EXTRA!
Design a flower bed for your school or home and work out the costs for it.

Consecutive numbers

Investigate adding two consecutive numbers together. Keep going until you reach 25.

1 + 2 = 3
2 + 3 = 5
3 + 4 =
4 + 5 =
+ =
+ =

+ =
+ =
+ =
+ =
+ =
+ =

Use a coloured pencil to record the answers on the 1–25 number grid by crossing through them.

1	2	3	4	5
6	7	8	9	10
11	12	13	14	15
16	17	18	19	20
21	22	23	24	25

Now investigate adding three consecutive numbers together. Record the answers on the 1–25 number grid. Use a different colour.

1 + 2 + 3 = 6
2 + 3 + 4 =
3 + 4 + 5 =
4 + 5 + 6 =
+ + =
+ + =
+ + =

Now try with four and five consecutive numbers. You may need to use a separate sheet of paper. Record your answers on the 1–36 number grid using different colours.

1	2	3	4	5	6
7	8	9	10	11	12
13	14	15	16	17	18
19	20	21	22	23	24
25	26	27	28	29	30
31	32	33	34	35	36

Are there any numbers not crossed out?

Which numbers were crossed out the most?

EXTRA!
Investigate adding consecutive numbers using a larger grid (maybe up to 100).

Number grid

1	2	3	4	5
6	7	8	9	10
11	12	13	14	15
16	17	18	19	20
21	22	23	24	25
26	27	28	29	30

Investigate ways of making each number in the grid by combining:

1 7 8 10 3

You can use +, −, x and ÷ and any of the numbers.
Don't use any number more than once each time.

For example, 23 and 27 can be made like this:

$10 \times 3 = 30 - 7 = 23$

$7 + 8 + 10 + 3 - 1 = 27$

See if you can make all the numbers

Can some numbers be made in more than one way?

EXTRA!
Choose five different starting numbers and try to make all the numbers in the grid.
You could change the grid to see if you can make all the new numbers.

Six digits

Choose four of these six digits.

2 3 5 6 7 9

Make two tens and units numbers, for example: 25 and 97 .

Investigate adding, subtracting and multiplying them. For example:

97	+	25	=	122
97	–	25	=	72
97	x	25	=	2,425

Reverse the digits and find out what happens.

79	+	52	=	131
79	–	52	=	27
79	x	52	=	4,108

Investigate what happens when you swap the pairs. For example, changing 25 and 97 to 57 and 29.

57	+	29	=	86
57	–	29	=	28
57	x	29	=	1,653

Now try reversing these digits as well.

75	+	92	=	167
75	–	92	=	-17
75	x	92	=	6,900

Which two digit numbers give you the highest and lowest additions, subtractions and multiplications? Compare your results with a friend's.

EXTRA!
Choose another set of four digits. Investigate them in the same ways.
You could use the back of the sheet.

Magic ladders

1st ladder	2nd ladder
0	1
2	3
4	5
6	7
8	9
10	11
12	13
14	15

Add any two numbers from the first ladder, for example, 4 + 8. Which ladder contains the answer? Try another two numbers. Does this always work?

Now try adding two numbers from the second ladder. Which ladder contains the answer? Why?

Tip: It may be helpful to use a colour code to record your findings on the ladders.

1st	2nd	3rd
0	1	2
3	4	5
6	7	8
9	10	11
12	13	14
15	16	17
18	19	20
21	22	23

Now try with three ladders. Complete the list below:

1st ladder + 1st ladder =
2nd ladder + 2nd ladder =
3rd ladder + 3rd ladder =
1st ladder + 2nd ladder =
1st ladder + 3rd ladder =
2nd ladder + 3rd ladder =

Investigate four ladders and complete the table.

1st	2nd	3rd	4th
0	1	2	3
4	5	6	7
8	9	10	11
12	13	14	15
16	17	18	19
20	21	22	23
24	25	26	27
28	29	30	31

+	1st ladder	2nd ladder	3rd ladder	4th ladder
1st ladder				
2nd ladder				
3rd ladder				
4th ladder				

Look for a pattern.

Dartboard

How many ways can you throw three darts and get a total that is a multiple of ten?

Do not count doubles, trebles or bull's-eyes.

Tip: Multiples of ten end in 0, for example: 10, 20, 30.

Now investigate making totals that are multiples of fives.

EXTRA!

Investigate making totals that are multiples of three.
Use doubles and trebles if you want.

Temperatures

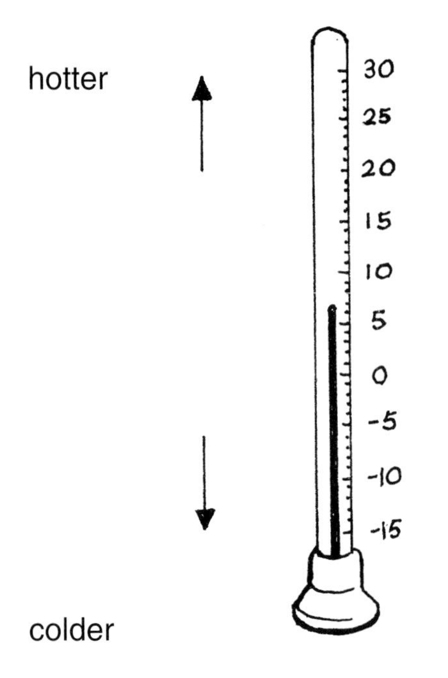

Average temperature in January

Cairo, Egypt	13°C
Montreal, Canada	-10°C
New York, USA	-1°C
Nairobi, Kenya	19°C
Buenos Aires, Argentina	23°C
Calcutta, India	20°C
Tokyo, Japan	3°C
Sydney, Australia	22°C
London, UK	4°C
Moscow, Russia	-13°C
Stockholm, Sweden	-3°C
Chicago, USA	-4°C
Anchorage, Alaska, USA	-11°C

If you flew from London to Tokyo in January is the temperature likely to get hotter or colder? How many degrees hotter or colder?

Tip: Use the picture of the thermometer to help you.

What if you flew from London to New York?

Which two cities have the greatest average temperature difference in January? How many degrees different?

Imagine you are a globe trotter. Choose a route round the world. Start and finish in London. Investigate the temperature changes you would be likely to experience.

Start city	Finish city	Start temperature	Finish temperature	Change (+/–)
London	Moscow	4°C	-13°C	-17°C
Moscow				
	London		4°C	

EXTRA!

Plot your imaginary journey on a world map.
Could you have taken a shorter route?

Fractions, 1

You will need the Fractions resource sheet, 1 (page 46), scissors, glue and coloured pencils. When dividing things into parts it is usually all right to describe any object that has been divided into two fairly equal parts as being in 'halves'. We sometimes even say, 'You take the bigger half'.

But in mathematics a half must be exact. An object must be divided exactly into two for the parts to be called 'halves'.

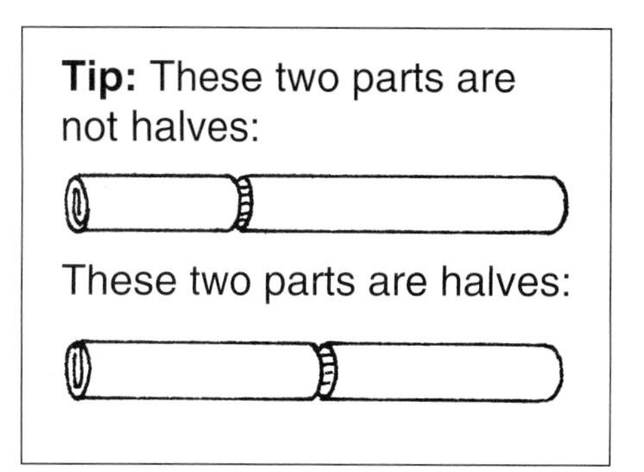

Tip: These two parts are not halves:
These two parts are halves:

Cut out the cloud and moon shapes on the resource sheet. Stick one of each shape here. Fold and cut the other shapes in half exactly. Shade each half a different colour. Lay one half on top of the other. Are they exactly the same size? Stick them here.

If an object is divided into four equal parts, each part is called 'a quarter'.

Cut out the flower and wall shapes on the resource sheet. Keep one whole and stick it here. Fold and cut the other shapes in half and shade each half a different colour. Stick them here. Fold and cut the last shape into four equal pieces. Shade each quarter a different colour. Compare the sizes of the quarters. Are they exactly the same size? Stick them here.

Compare the size of two quarters with the halves. What do you notice?

EXTRA!

Find out what each piece is called if an object is divided exactly into three pieces, five pieces, eight pieces and ten pieces.

Fractions, 2

There are 8 cubes in my set. One cube is black. This is $\frac{1}{8}$ of my set.

Seven cubes are white. This is $\frac{7}{8}$ of my set.

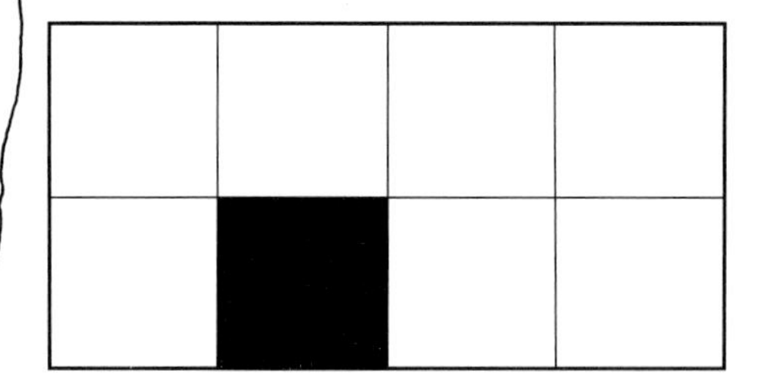

Collect 16 interlocking cubes of two colours. How many different numbers of each colour can be used to make a set of 8?

Colour in the squares to show your sets and write the fractions below.

colour 1 colour 2

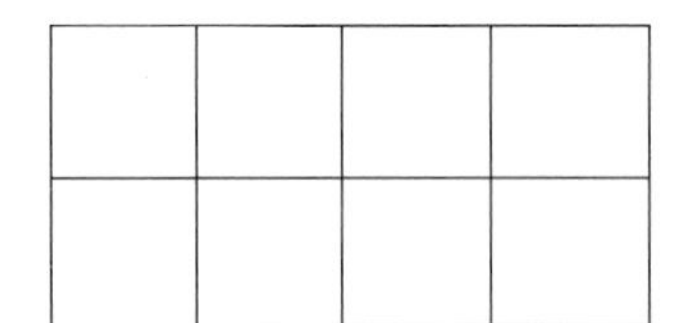

colour 1 colour 2

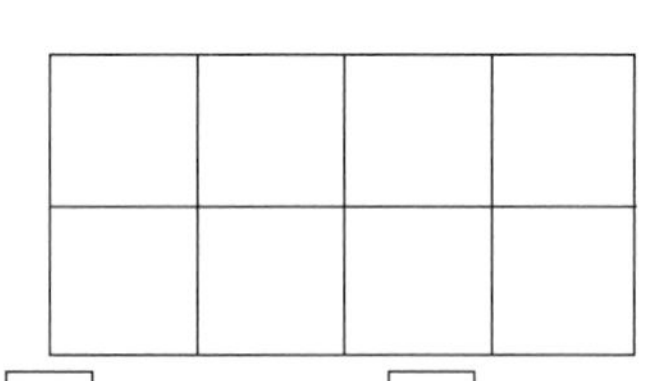

colour 1 colour 2

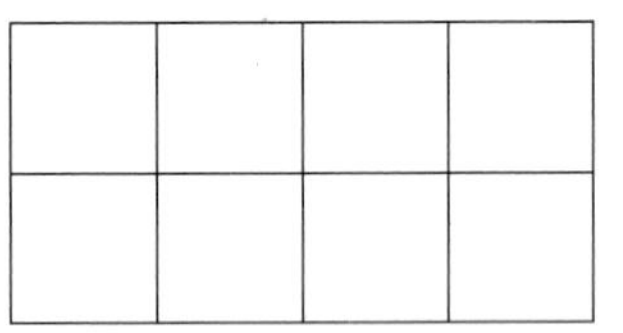

colour 1 colour 2

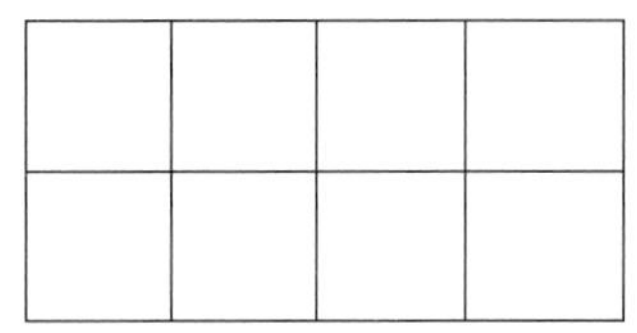

colour 1 colour 2

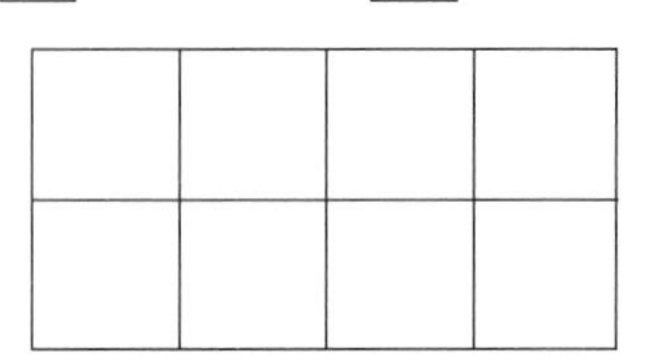

colour 1 colour 2

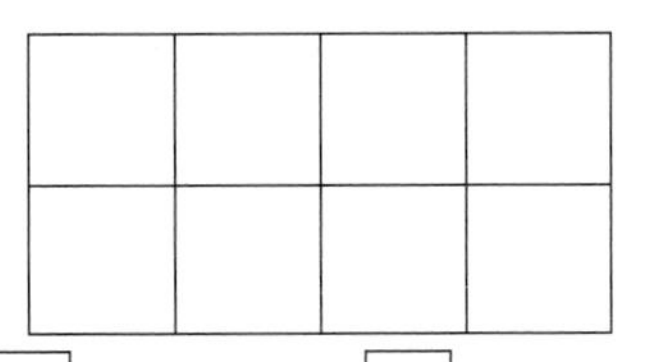

colour 1 colour 2

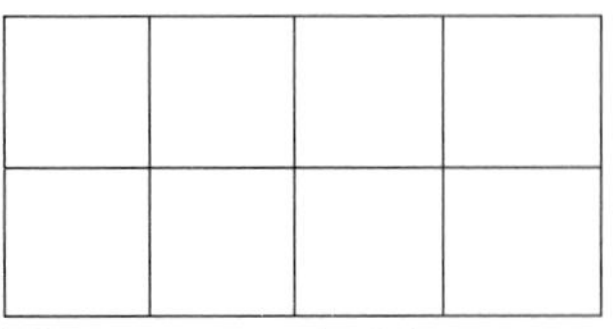

colour 1 colour 2

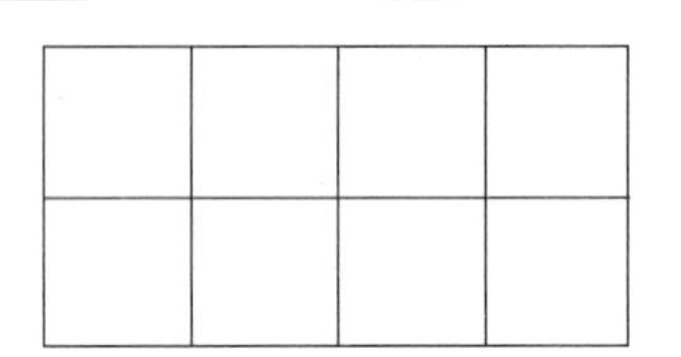

colour 1 colour 2

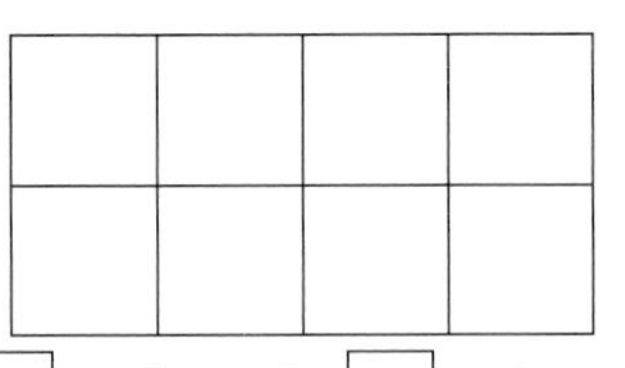

colour 1 colour 2

colour 1 colour 2

colour 1 colour 2

EXTRA!
Investigate ways of dividing a set of 12 cubes using two, three or four different colours.

Sharing

Two friends want to share three apples between them.

One way round this to have one whole apple each and divide the last apple into half.
Each friend would have:

$$1 \quad + \quad \tfrac{1}{2} \quad \longrightarrow \quad 1\tfrac{1}{2} \text{ apples}$$

But, the apples are all different sizes and they want to be fair. They have a good idea: divide up each apple into half and each friend can have half of every apple.

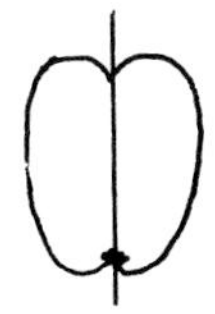 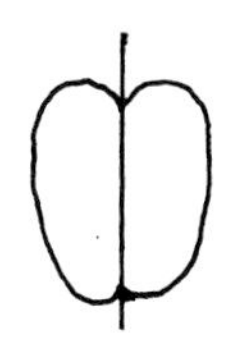 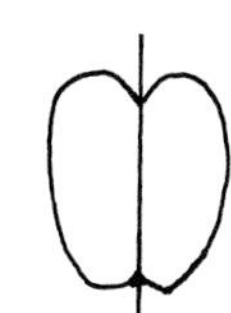

Each friend has:

$$3 \quad \times \quad \tfrac{1}{2} \text{ apples} \quad \longrightarrow \quad \tfrac{3}{2} \text{ apples}$$

Can you think of a way to share these apples amongst these children?

    and   

   and 

    and 

EXTRA!
Investigate ways of sharing sets of 12 between different numbers of children.

Fraction patterns

There are two different ways of dividing a square in half using a straight line.

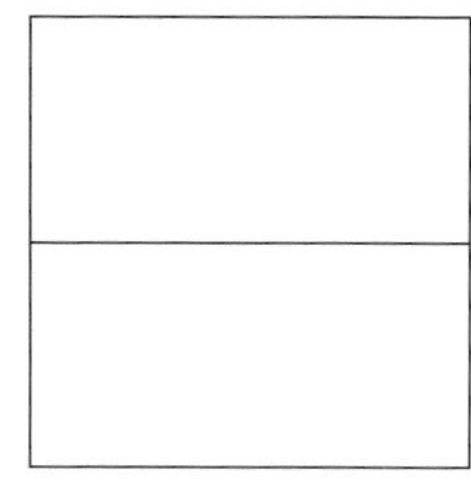

side to side

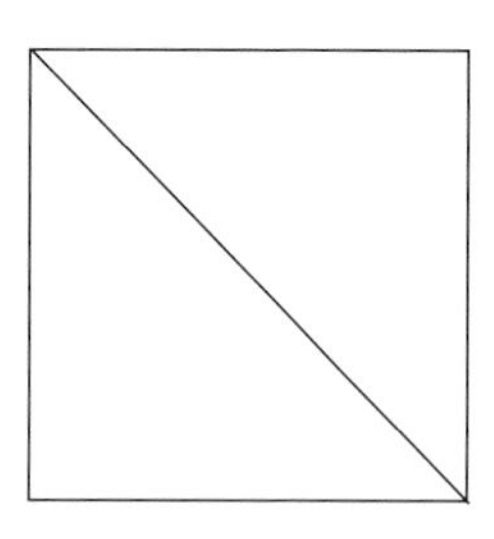

corner to corner

Use the squares below to investigate the number of ways to divide a square into quarters using straight lines.

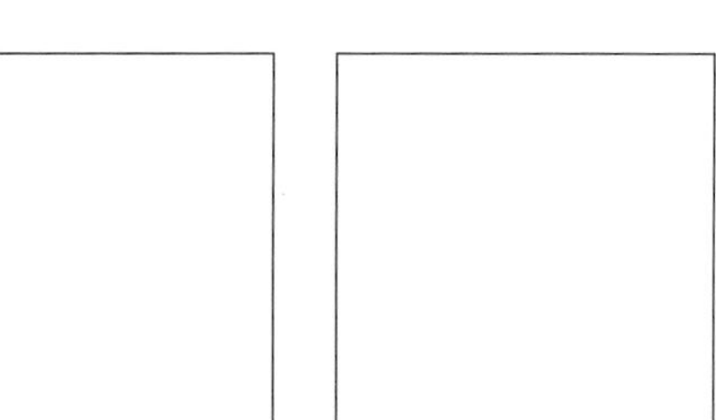

There are probably fewer ways to make a quarter than you think.

Choose either a square divided in half or quarters and investigate making a repeating pattern of squares in this grid.

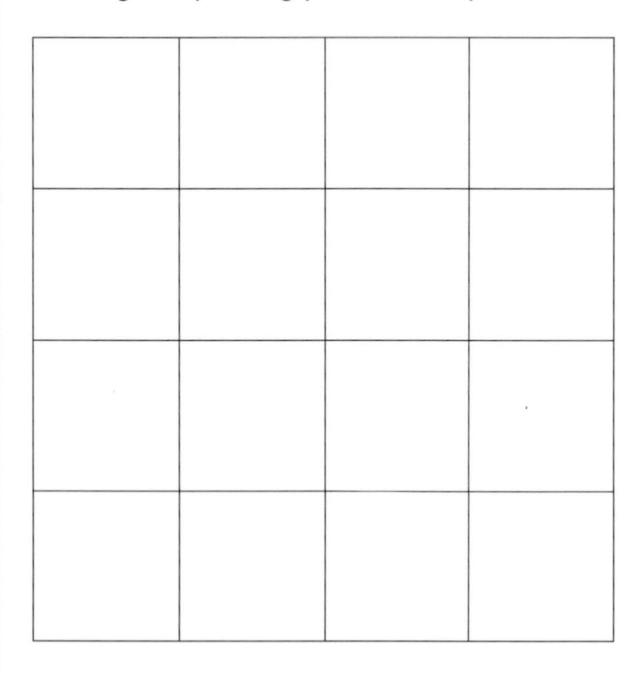

EXTRA!
Use the Fractions resource sheet, 2 (page 47) to investigate making more repeating patterns of halves or quarters. For example:

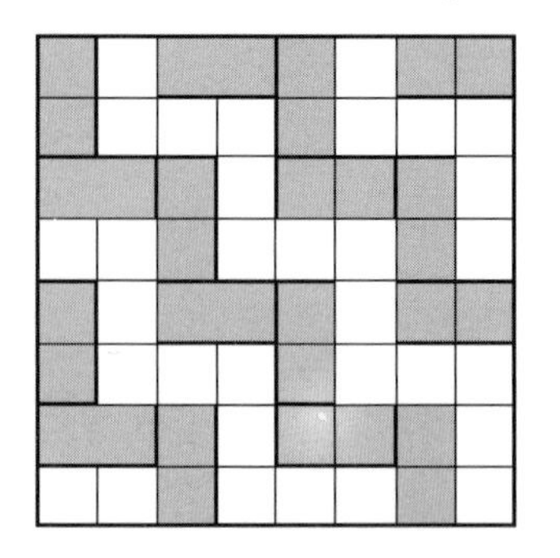

Making one

You will need the Fractions resource sheet, 3 (page 48).

Cut out the fraction pieces.

Place the **'one'** piece on the table and lay a $\frac{1}{2}$ piece below it.

one	
$\frac{1}{2}$	

Investigate different ways of making the $\frac{1}{2}$ up to a whole **'one'** using one or more of the other fraction pieces.

Tip: There are four ways of making one using $\frac{1}{2}$ and one other type of fraction, for example: $\frac{1}{4}$ or $\frac{1}{6}$.

How many different ways can you make **'one'** using $\frac{1}{2}$ and two other other types of fraction?

EXTRA!
Investigate how many ways you can make **'one'**,
starting with $\frac{1}{3}$ or $\frac{1}{4}$.

Twelfths

You will need the Fractions resource sheet, 3 (page 48).

Cut out the individual pieces.

Investigate different ways of making a half.

$\frac{1}{2}$ = ☐ quarters

$\frac{1}{2}$ = ☐ sixths

$\frac{1}{2}$ = ☐ twelfths

Tip: To keep the pieces from blowing away you could use small bits of Blu-Tack.

Investigate ways of making a quarter and a third using only one type of fraction.

Make:

- $\frac{3}{4}$ using twelfths
- $\frac{2}{3}$ using sixths
- $\frac{2}{3}$ using twelfths
- $\frac{5}{6}$ using twelfths
- $\frac{9}{12}$ using quarters
- $\frac{4}{6}$ using thirds

You might like to colour each fraction strip a different shade.

EXTRA!
Investigate ways of making a half using more than one type of fraction.

Fraction chains

Remember: $\frac{1}{2} = \frac{2}{4}$ one half = two quarters

$\frac{2}{4} = \frac{4}{8}$ two quarters = four eighths

So $\frac{1}{2} = \frac{2}{4} = \frac{4}{8}$

Look at the pattern in the top numbers: $1 \rightarrow 2 \rightarrow 4$

They have been multiplied by 2.

Now look at the pattern in the bottom numbers: $2 \rightarrow 4 \rightarrow 8$

They have been multiplied by 2 as well.

Continue the pattern:

$$\frac{1}{2} = \frac{2}{4} = \frac{4}{8} = \frac{\quad}{16} = \frac{\quad}{\quad} = \frac{\quad}{\quad} = \frac{64}{128}$$

What happens if we multiply by 3?

$\frac{1}{2} = \frac{3}{6}$

$\frac{3}{6} = \frac{9}{18}$

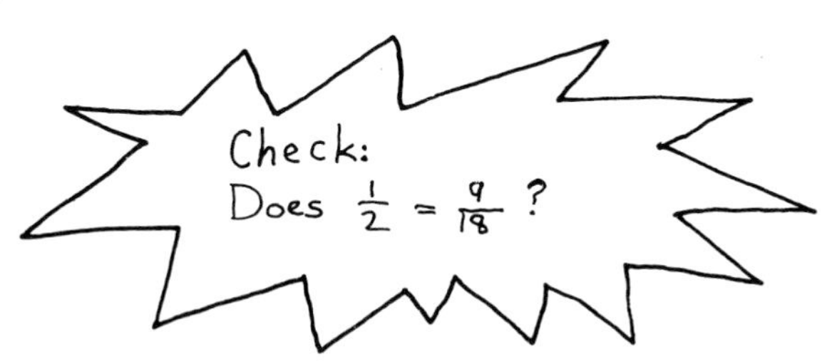
Check:
Does $\frac{1}{2} = \frac{9}{18}$?

Continue the pattern using a calculator:

$$\frac{1}{2} = \frac{3}{6} = \frac{9}{18} = \frac{\quad}{54} = \frac{\quad}{\quad} = \frac{\quad}{\quad} = \frac{729}{1458}$$

EXTRA!
Choose a different starting fraction and make up some fraction chains by multiplying the top and bottom number by the same whole number.

Matching pairs

You will need to work with a friend. Cut out the cards on your sheet.

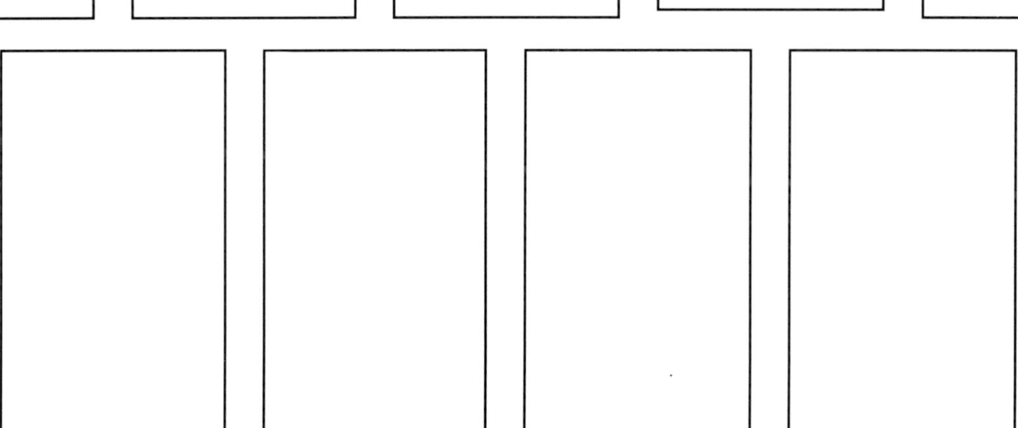

Use your calculator to work out the equivalent decimal fractions for these vulgar fractions:

$\frac{1}{2}$ =	$\frac{1}{5}$ =	$\frac{2}{5}$ =
$\frac{1}{10}$ =	$\frac{3}{10}$ =	$\frac{7}{10}$ =
$\frac{9}{10}$ =	$\frac{3}{5}$ =	$\frac{4}{5}$ =

Tip: To find the decimal fraction enter the top number of the fraction and divide by the bottom number. For example: $\frac{3}{10}$ = 0.3 (3 ÷ 10 = 0.3).

One of you should write the vulgar fractions on your set of cards. The other should write the decimal fractions.

Use the 18 cards to play a matching pairs game. Lay the cards face down on the table. Take it in turns to turn over two cards. If they make a matching pair you can keep them and continue playing. If they do not match, turn them back over and the other player has a turn. The winner is the person who has collected the most cards.

EXTRA!

Add cards for $\frac{1}{4}$, $\frac{3}{4}$, $\frac{3}{8}$, $\frac{5}{8}$, $\frac{7}{8}$ and their decimal equivalents and play the matching pairs game again.

Find the route

Find the route from Sam's house to the school by matching the fractions which are the same.
Then find the route from Sally's house to the swimming pool.

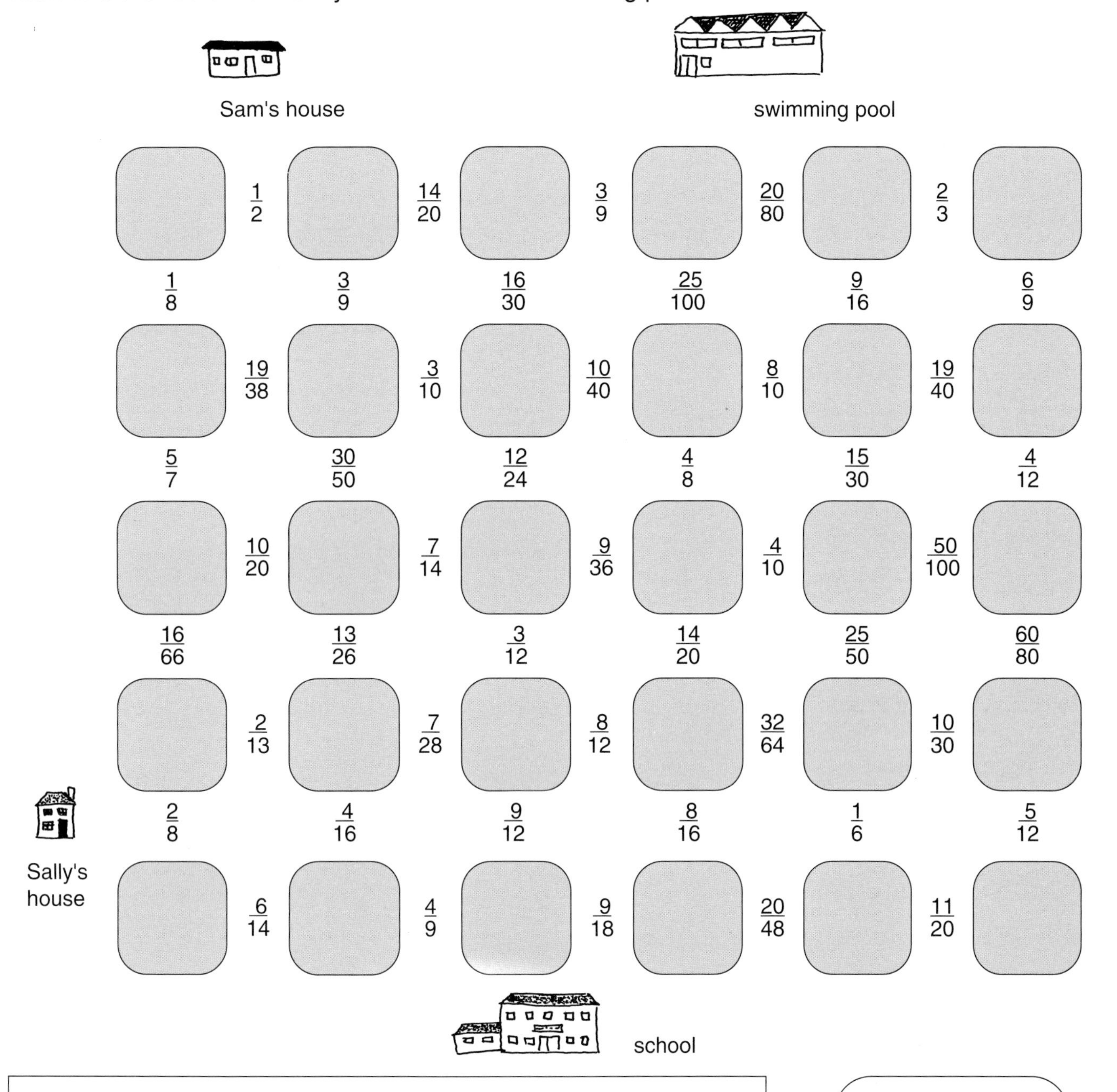

Tip: Use your calculator to help you. Enter the top number then press divide. Enter the bottom number then press equals.

All fractions that are the same have the same decimal. For example: $\frac{1}{2}$ is the same as $\frac{4}{8}$.

On the calculator $1 \div 2 = 0.5$

and $4 \div 8 = 0.5$

They are called equivalent fractions.

EXTRA!
Make up your own maze using matching fractions.

Surveys

A group of six friends take a survey of their favourite games.
This is the result:

Game	I have played	I have been to a game	I have watched it on TV	Don't know
Tennis	~~IIII~~	–	IIII	I
Football	~~IIII~~	IIII	~~IIII~~ I	–
Rounders	II	–	–	IIII
Cricket	I	I	IIII	–
Netball	III	I	I	II
Hockey	–	I	I	IIII
Badminton	II	II	I	II
Ice hockey	–	I	III	I
American football	–	I	III	II
Rugby	–	II	III	II

What fraction of games have been played by the friends? Count how many games there are altogether. This is the bottom number. Count how many games have been played. This is the top number. $\frac{6}{10}$ of games have been played by the friends.

On another piece of paper make up some other questions and statements about the games. For example:

> There are 6 friends and 5 have played tennis. So $\frac{5}{6}$ of the group have played tennis.
>
> What fraction of the group had played netball?

EXTRA!
Carry out a survey of your own with a small group of friends.
Look at the results and use fractions to help you describe them.

Buying sweets

In the sweet shop, nut crunch costs 40p for 100g. Complete the table below to find the cost of 200g, 300g and 400g of nut crunch.

	100g	200g	300g	400g
nut crunch	40p			

The graph below shows the cost of four different sweets. Use the information in the table to plot a line to show the cost of nut crunch.

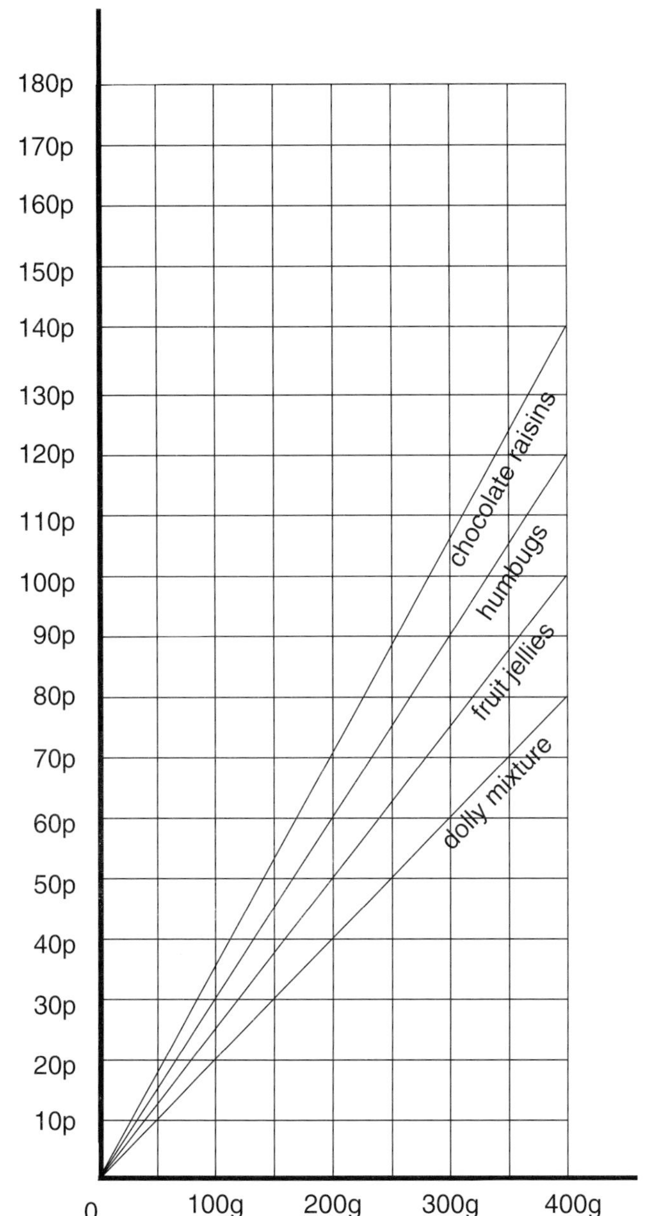

How much does 200g of dolly mixture cost?

How much does 300g of humbugs cost?

How much does 100g of fruit jellies cost?

Which is the most expensive sweet?

What weight of humbugs could be bought for 60p?

If you had £1.00 to spend on sweets in this shop, what would you buy?

EXTRA!
Make up five more questions about the graph. Use a separate sheet of paper.

Climate

Look at these temperatures. What can you say about the climate in each place through the year? Is it hot or cold in comparison to the other places? How much does it vary?

Average temperatures in °C													Mean temperature	Range of temperature
	Jan	Feb	Mar	Apr	May	June	July	Aug	Sept	Oct	Nov	Dec		
Quito, Ecuador	15	15	15	15	15	14	14	15	15	15	15	15		
Colombo, Sri Lanka	26	26	27	28	28	27	27	27	27	27	26	26		
New Delhi, India	14	17	22	28	33	34	31	30	29	26	20	15		
Verkhoyansk, Russia	-49	-44	-30	-13	2	12	15	11	3	-14	-36	-46		

Source: Philip's Modern School Atlas

Calculate the **mean** temperature for each place for the year. Places can have very similar mean temperatures, but this doesn't mean that they will have the same temperature throughout the year!

The **range** of temperatures over the year tells us much more about the climate. Calculate the range of temperature for each place.

Tip: To calculate the **mean**, total the 12 temperatures on a calculator and divide by 12. Round the answer to the nearest whole number.

To calculate the **range** of temperatures, find the difference between the lowest and the highest temperature. For example: -23°C and 17°C. The range is 40°C.

EXTRA!

Use an atlas to find some places with large and small temperature ranges. Compare the mean temperatures for the year for these places.

Digits, 1

In our place value system the position of a digit tells you its value.

Take, for example, the number **543**:

The **5** shows how many hundreds there are:	500	=	5	x	100
The **4** shows how many tens there are:	40	=	4	x	10
The **3** shows how many units there are:	3	=	3	x	1

Rearrange these digits to make as many different numbers as you can:

3 1 2

Put your numbers in order beginning with the smallest.

Now do the same thing with these digits:

4 0 5

Think about what difference the zero makes.

EXTRA!

Make the largest and smallest numbers you can, using 7, 2, 1 and 4.

Try with a different set of four digits.

Multiples of ten

Reminder:

59	x	10	=	590
17	x	10	=	170
64	x	10	=	640
123	x	10	=	1,230

When multiplying by 20, it is helpful to remember that 20 = 2 x 10. You can either multiply the number by 2 first and the answer by 10, *or* multiply by 10 first and the answer by 2.

For example:

34 x 20 =
34 x 2 x 10 =
68 x 10 = **680**

or

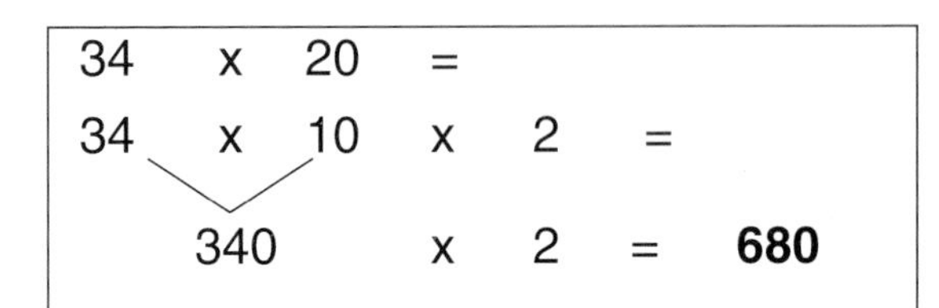

The same methods work for other multiples of 10, such as 30, 40, 50, 60, 70, 80 and 90.

For example:

21 x 30 =
21 x 3 x 10 =
63 x 10 = **630**

or

21 x 30 =
21 x 10 x 3 =
210 x 3 = **630**

12 x 50 =
12 x 5 x 10 =
60 x 10 = **600**

or

12 x 50 =
12 x 10 x 5 =
120 x 5 = **600**

Try multiplying these using both methods:

32 x 40

18 x 60

42 x 70

53 x 80

65 x 90

16 x 50

Use the back of the sheet if you need more room.

EXTRA!

You can use the same methods to multiply by multiples of 100, such as 200, 300 and 400. What do you think the methods might be? Make up some multiplication sums using multiples of 100 and try your methods.

High numbers

You will need two copies of the Numbers resource sheet (page 45) and a friend.

Cut out the number cards and keep the cards between 0 and 9.

Shuffle them and put them in a pile face down.

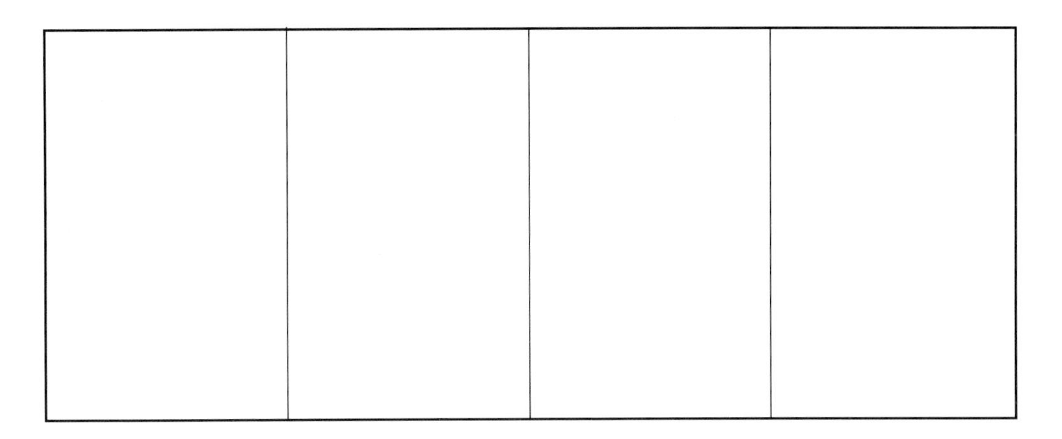

Work with one grid each. Don't let your friend see your grid.

In turns, take a number card and place it anywhere on your grid. Try to make the highest number possible, but do not move the cards once you have put them down.

When your grids are complete, compare numbers. The person with the highest number wins.

Rearrange the four cards to try to make an even higher number.

EXTRA!
Try to make the lowest number possible.
Then make numbers that are nearest to 2,000, then 5,000.

Cross numbers

Across

1 One Thousand and One
3 Ninety Nine
5 One Thousand, Two Hundred and Thirty Four
7 Ten
8 Fifty Thousand and Four Hundred
10 One Hundred
11 Three Hundred and Four
12 One Thousand, Five Hundred and Two
14 One Million, Four Hundred and Seventy Two Thousand, and Nine
17 Twenty Nine Thousand and Ninety Nine
18 Fifty Thousand, Seven Hundred and Three

Down

1 Eleven Thousand, One Hundred and Eleven
2 One Thousand and Forty
4 Ninety Thousand and Four
6 Three Million, Five Hundred and One Thousand, Seven Hundred and Five
7 Ten Thousand and Twenty
9 Four Hundred and Thirty Thousand and Ninety Seven
13 Five Thousand, Two Hundred and Ninety
14 One Hundred and Twenty
15 Forty Nine
16 Nine Hundred and Three

EXTRA!

Make up different clues to the same answers in the cross number, using +, –, x and ÷. You might like to use a calculator.

Rounding...

Rounding is a way of making numbers more simple to use. You can round to the nearest 10, 100, 1,000 or even 1,000,000. This is how you round to the nearest 10.

Look at this part of a number line.

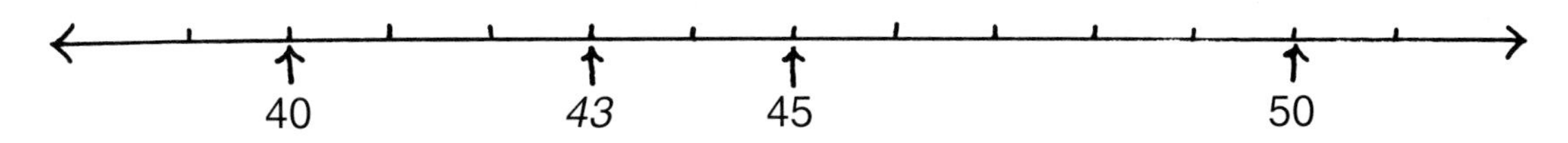

43 is nearer to the 40 mark than the 50 mark, so it is rounded down to 40. 43 rounded to the nearest 10 is 40.

What is 49, rounded to the nearest 10?

45 presents a problem. It is exactly half way between 40 and 50. Mathematicians agree to round it up to 50.

You can do the same thing for 100s and 1,000s.

Round these distances to the nearest 10 km.

54 km
78 km
125 km
99 km
60 km
213 km
187 km

Round these heights of mountains to the nearest 100 m.

Snowdon	1,085 m
Ben Nevis	1,344 m
Everest	8,848 m
Mont Blanc	4,807 m
Mount Fuji	3,192 m
K2	8,611 m
Mount Whitney	4,418 m
Kilimanjaro	5,895 m

EXTRA!

Round these football crowds to the nearest 1,000 people.

Sheffield Wednesday	vs	Blackburn	24,699
Chelsea	vs	Wolverhampton	14,408
Liverpool	vs	Everton	29,340
Sheffield Utd	vs	Leeds	19,425

Population

The table shows ten countries, their capital cities and the population (number of people who live there) rounded to the nearest 1,000.

Country	Population	Capital city	Population	Percentage
Austria	7,583,000	Vienna	1,540,000	
Belgium	9,845,000	Brussels	1,331,000	
France	56,138,000	Paris	9,319,000	
Iceland	253,000	Reykjavik	143,000	
Japan	123,460,000	Tokyo	11,936,000	
Kenya	24,031,000	Nairobi	1,429,000	
Kuwait	1,500,000	Kuwait	189,000	
Singapore	2,700,000	Singapore	2,334,000	
Turkey	58,687,000	Ankara	3,022,000	
United Kingdom	57,237,000	London	6,378,000	

Source: Philip's Geographical Digest 1994-95

Calculate the percentage of the country's population which lives in the capital city.

Tip: Enter the capital city's population in your calculator. Divide it by the country's population. Multiply by 100. Round to the nearest whole number.

For example:
Spain: 39,187,000 Madrid 3,121,000

Enter 3,121,000 and divide by 39,187,000.
Multiply by 100. The calculator shows 7.96437. This is 8% rounded to the nearest whole number.

Which capital city has the highest percentage of the country's population?

Which capital city has the lowest percentage of the country's population?

EXTRA!
Investigate the percentage of the population that lives in the five largest cities in the United Kingdom.

Digits, 2

In our place value system the position of a digit tells you its value.

Complete these patterns. Start with the digit in the star.

Tip: Don't forget to write in all the zeros.

x100	x10	x1	x0.1	x0.01	x0.001
500	50	5	0.5	0.05	0.005
		7	0.7		
		3			
400				0.04	

Try making larger and smaller numbers using a different digit in the star.
Use a calculator to help you.

EXTRA!
Try with these. Use the calculator.

	57				
125					

Which makes the smallest number?
Which makes the largest number?

Decimal fractions, 1

One hundred pennies are equivalent to one pound coin (£1.00).

Ten 10p coins are also equivalent to one pound coin (£1.00).

Ten pennies are equivalent to ten pence (10p).

You can write 1p in pounds like this: £0.01.

Enter 0.01 into your calculator.

Multiply by 10.
The calculator says 0.1.
In pounds this means £0.10 or 10p.

Multiply by 10 again.
The calculator says 1. This means £1.

Multiply by 10 again.
What does the calculator say?

What does it mean in pounds?

Keep going until you reach £1,000,000.
Record all your results. How many times did you multiply by 10?

Enter amount

Multiply by 10

Record the result in pounds.

Continue until the amount is over £1,000,000

Show how you would write 3p, 5p, 10p and 20p using the pound sign. Then follow the flow chart for each one in turn. Write down what the calculator shows at each stage on a separate sheet of paper.

EXTRA!
Change the flow chart so that it divides by 10.
Start from £1,000,000 and continue until you reach 1p.

Decimal fractions, 2

Look at the pattern in the box. Continue it for two more stages.

5	x	100	=	500
5	x	10	=	50
5	x	1	=	5
5	x	0.1	=	0.5
5	x	0.01	=	0.05
			=	
			=	

Investigate the pattern using other starting numbers that are below 10.

Now complete the pattern in this box.

15	x	100	=	1,500
15	x	10	=	150
15	x	1	=	15
15	x	0.1	=	0.15
15	x	0.01	=	0.015
			=	
			=	

Investigate the pattern using other starting numbers that are above 10.

EXTRA!
Investigate what pattern you get when you use starting numbers above 100.

Buying stationery

Use your calculator to find the cost of buying 1, 10, 100, 1,000 and 10,000 of each item of stationery.

Item	1	10	100	1,000	10,000
rubber	£0.09				
pencil	£0.06				
pen		£0.50			
sharpener			£2.00		
ruler				£300.00	
protractor			£17.00		
compass	£0.25				
coloured pencil		£1.10			
felt-tip pen					£1,500.00

Tip: Enter the amounts into your calculator in pounds. If you don't, the numbers will get so big you might run out of space.

EXTRA!

You could get a discount if you buy lots of one item. Calculate the prices for the above items if you get:

- 10% discount for buying 100 items,
- 15% discount for buying 1,000 items and
- 20% discount for buying 10,000 items.

If you need help calculating percentages ask your teacher for a copy of page 36.

Decimal units

Metric measurements are based on the decimal system. Complete the boxes below.

1 metre	=	10 decimetres	=	100 centimetres	=	1,000 millimetres
1 litre	=	[ ] decilitres	=	[ ] centilitres	=	[ ] millilitres

Use the information above to help you find these relationships:

500 mm = [ ] cm

[ ] cm = 4 dm

3 dm = [ ] mm

4 m = [ ] cm

10 m = [ ] mm

Tip: The measurements are written using symbols:

metre	=	m
decimetre	=	dm
centimetre	=	cm
millimetre	=	mm

Make up five relationships of your own, using the litre family. Use your own numbers.

What symbols are used for these measurements?

EXTRA!

Metric measurement uses interesting names to show very large and very small measurements. For example: a microsecond is one millionth of a second and a kilometre is 1,000 metres.

Find some other examples in a reference book. Which is the largest you can find? Which is the smallest?

Mixed fruit

Make up different bags of fruit using the fruit in the picture below. Each bag must weigh as close as possible to 1 kg.

Lemons
150 g
5p each

Apples
125 g
10p each

Bananas
75 g
20p each

Oranges
175 g
25p each

Grapefruit
200 g
35p each

Pears
100 g
15p each

EXTRA!
Make up mixed bags of fruit that cost exactly £1.00.

Baggage allowance

Passengers travelling on some airlines are only allowed 20 kg of baggage allowance.
They can have as many bags as they like but the total weight must not be more than 20 kg.

Find out which sets of baggage are within the limit.

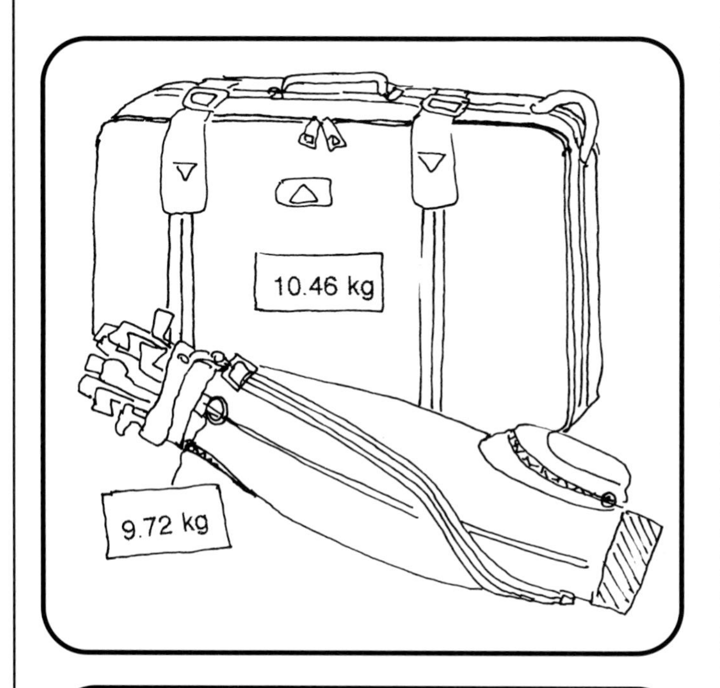

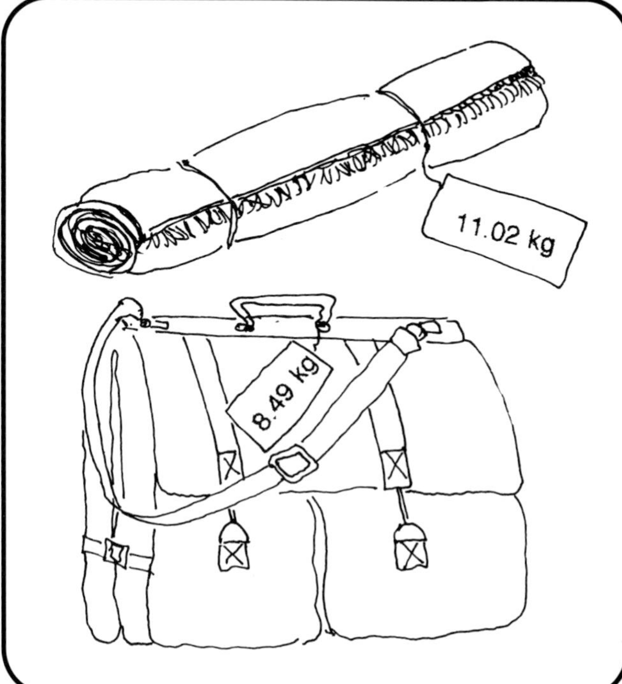

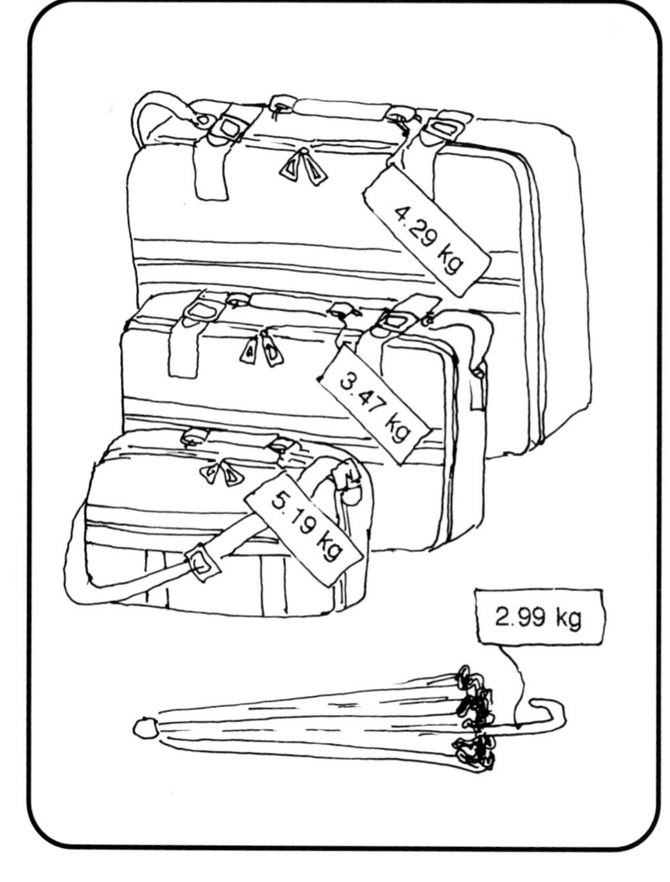

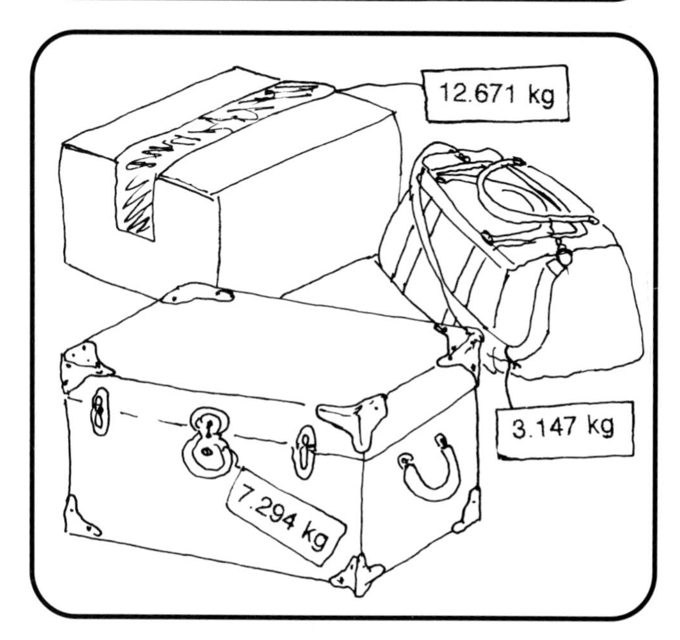

EXTRA!
Find three objects that weigh about 1 kg in total. How close to 1 kg can you get?

Self-assessment sheet

Name ______________________________

I can …	Date
investigate different number systems	
use the addition facts to 20	
investigate the 2, 5 and 10 times tables	
use the 2, 5 and 10 times tables	
divide amounts of money	
calculate the total cost of several items	
investigate the properties of numbers	
combine numbers using +, –, x and ÷	
investigate tens and units	
record my work systematically	
investigate number bonds	
calculate with negative numbers	
identify halves and quarters	
calculate half of a set	
calculate using fractions of whole units	
investigate fractions of shapes	
identify equivalent fractions	
investigate ways of making a whole unit using fractions	
investigate equivalent fractions	
calculate decimal fractions from vulgar fractions	
use fractions to describe information	
calculate and use the mean and range of sets of data	
recognize the importance of the position of a digit	
multiply with multiples of ten	
read numbers in words	
round to the nearest 10, 100 and 1,000	
calculate percentages of amounts	
use decimal fractions	
follow instructions to solve a problem	
investigate patterns in decimal fractions	
use decimal notation of money	
investigate equivalents in decimal units of measurement	
use decimal units of measurement	

Numbers resource sheet

6	13	20
5	12	19
4	11	18
3	10	17
2	9	16
1	8	15
0	7	14

Fractions resource sheet, 1

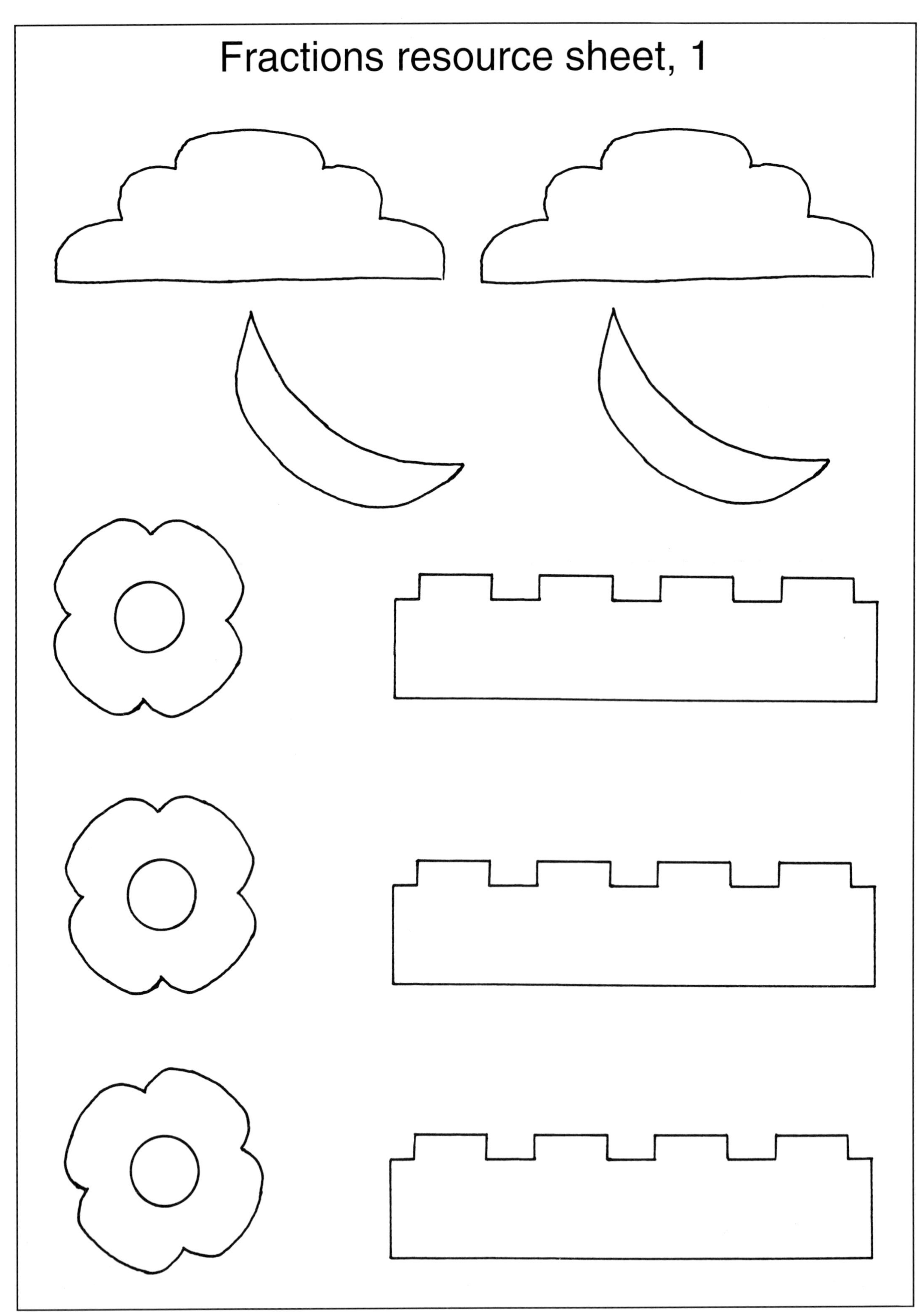

Fractions resource sheet, 2

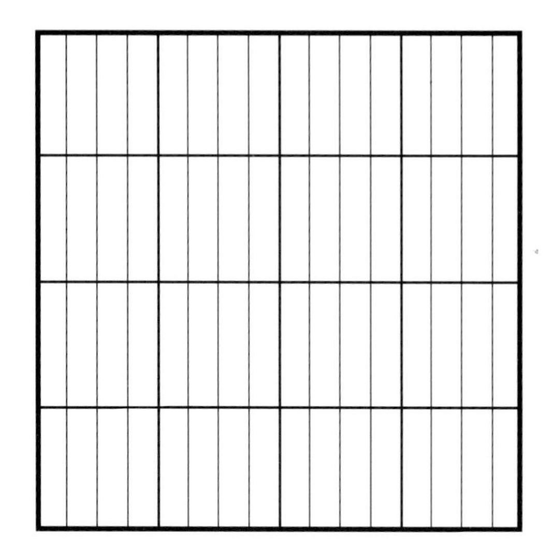

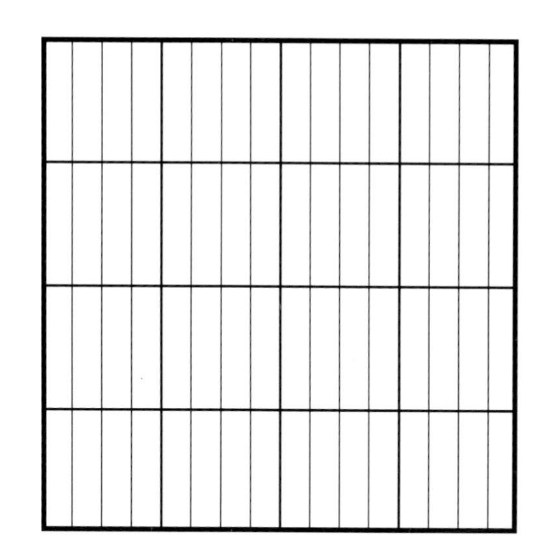

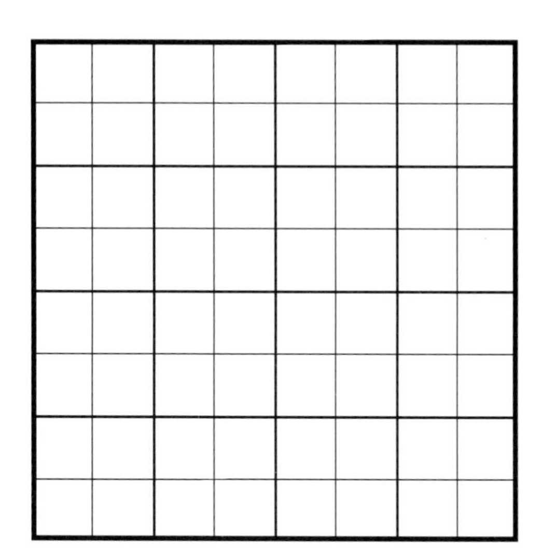

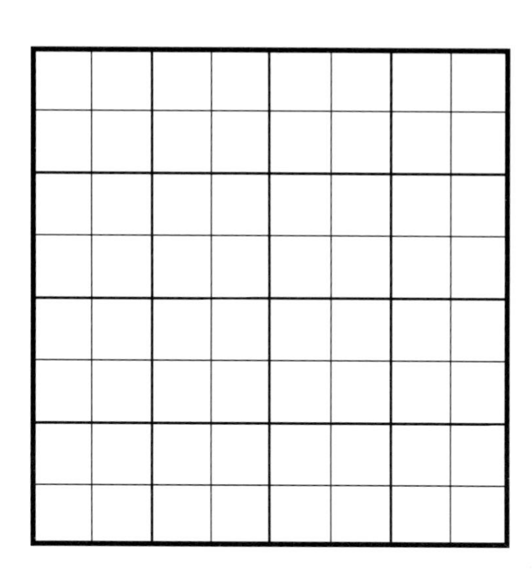

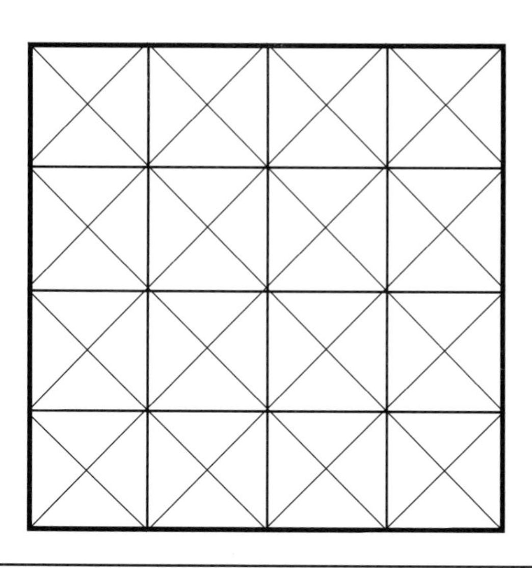

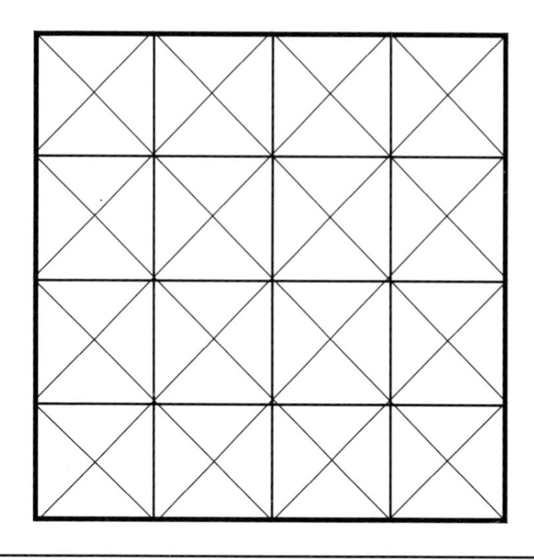

Fractions resource sheet, 3

one											
$\frac{1}{2}$						$\frac{1}{2}$					
$\frac{1}{3}$				$\frac{1}{3}$				$\frac{1}{3}$			
$\frac{1}{4}$			$\frac{1}{4}$			$\frac{1}{4}$			$\frac{1}{4}$		
$\frac{1}{6}$		$\frac{1}{6}$		$\frac{1}{6}$		$\frac{1}{6}$		$\frac{1}{6}$		$\frac{1}{6}$	
$\frac{1}{12}$	$\frac{1}{12}$	$\frac{1}{12}$	$\frac{1}{12}$	$\frac{1}{12}$	$\frac{1}{12}$	$\frac{1}{12}$	$\frac{1}{12}$	$\frac{1}{12}$	$\frac{1}{12}$	$\frac{1}{12}$	$\frac{1}{12}$